TROPICAL
BROILER CHICKEN
MANAGEMENT GUIDE

TROPICAL BROILER CHICKEN MANAGEMENT GUIDE

Hauhouot Diambra-Odi

To order additional copies of this book, contact:
Xlibris LLC
1-888-795-4274
www.Xlibris.com
Orders@Xlibris.com
140842

Dedication

This booklet is dedicated to my mother, Koussoh Maria Diambra-Odi, for her love and dedication to her children, and to the Lord for all His blessings.

Acknowledgments

I would like to thank Dr. Lee Yudin, dean of the College of Natural and Applied Sciences and my colleagues Dr. Mari Marutani and Dr. Jim McConnell for their support and Dr. Tim Guile for reviewing the manuscript. I am grateful to my wife, Marcelle and children Odi Jr., Foushi and Jacques for their continued love.

Foreword

Writing a book on Tropical Broiler Chicken Management Guide involves time, dedication, knowledge and a great deal of persistence. The author's interest in chickens is more than a hobby but an admiration for these feathered friends.

I have known the author and his family for nearly 20 years and have been extremely impressed by his professional skills not only as a scientist but also as an educator of both formal and informal classroom settings. To my surprise, I have learned that the author has another keen interest besides Avialae and that is watching sunsets. One might think these are two very different entities but if one thinks of the multi-colored sunsets that we witness on our islands and the diversity of feather coloration then it is possible to see the connections between fowl and sunsets.

The information contained in this book allows both the backyard entrepreneur and commercial broiler grower the skills necessary to make a profit and gain a complete understanding in what it takes to produce boilers for market enterprises. The author pays special attention to tropical conditions like those that exist in Guam and in Micronesia. In Micronesia, hot and humid conditions are prevalent year around and tropical disturbances are the norm rather than the exception. This means that not only raising healthy stock in structures that need ideal ventilation but nutritional factors can greatly influence possible diseases common in tropical broiler production. The author continues to use practical examples of how important the right feed is, an area of his expertise, stresses the awareness of environmental standards, waste management, and the importance of sanitation in a biological system.

I think each reader will be able to gain a broad insight into the complete management guide of tropical broiler chickens. The book contains both general and specific information that will allow the entrepreneur to make important knowledge based decisions for a productive enterprise. This book is a must for anyone who is interested in the management of tropical broiler chickens.

Lee S. Yudin
Dean, College of Natural and Applied Sciences

Table of Contents

Introduction

An efficient management program is the key to a profitable and satisfying poultry enterprise. This brochure is designed to assist the **broiler** grower in obtaining optimum performance from the broiler flock.

Brooding and Rearing Chicks

It is important to remember that chicks must have a good growing environment (temperature, humidity, feeder and waterer density, **floor density**, ventilation, adequate feed and clean water) in order to reach their full genetic potential.

The following 10 points need to be considered:

1. Day-old chicks should come from a hatchery with a good reputation because a chick's quality (size, uniformity, cleanliness, alertness, freedom from deformities, etc.) is the guarantee for a profitable grow-out operation.
2. Chicks can be ordered by mail and can survive a two-day journey without adverse effects on their future performance.
3. Before arrival (three to four hours prior) the **brooder** should be checked to be sure it is clean and dry, provided with fresh, clean litter and water warmed to 34-35 C. Upon arrival chicks need to be placed immediately in the brooder.
4. It is better to use an **all-in/all-out** program. This is a management practice whereby birds of the same age are raised and sold all at once.

5. Different types of brooders are available, using either propane gas or electricity. However, 60 to 100 watt light bulbs are appropriate for providing heat to 50 chicks, when placed at 30 to 50 cm above the chicks. Improper brooding temperature can produce a condition known as **pasting**.

6. Install cardboard or newspaper around the brooder guard to protect the chicks from draft and to keep the chicks close to the heat source.

7. Several types of chick feeder and waterer are available on the market. However, egg flats make good feeders for day-old chicks until they reach seven days of age. Large commercial operations use automatic feeders and waterers.

8. Feeders and waterers should be arranged in such a way that chicks can find feed and water without having to walk too far.

9. Watch for feed wastage and avoid having the feeders overflow. It is always better to provide a large number of half-full feeders than to fill up fewer feeders to the point of overflow. One egg flat will provide sufficient feed for 20 day-old chicks until they are seven days old.

10. Observe the chicks' behavior over the next two to three hours.

Troubleshooting in the Brooder

Several problems related to temperature that need to be watched for in the brooder are:

- Improper brooding temperature is indicated by an uneven distribution of chicks in the brooding area.
- If chicks crowd around the heat source, the brooder is not warm enough.
- If chicks try to escape the heat and pile along the brooder guard, the brooder is too warm.
- If chicks pack on one side, drafts are present in the brooding area.

Housing

<u>The housing for poultry may be one of the following types:</u>

- Open sheds
- Semi-enclosed
- Enclosed house (light and temperature control)
- Free range (new trend for organic farms)

On Guam and Micronesia, the grower has to plan for securing the house during typhoon conditions and must be able to store water for at least one week's consumption. Under the tropics in general, open sheds are the preferred housing for broiler chickens.

<u>Space Requirements:</u>

The space requirement is 10 birds per square meter of floor space. Free range will provide additional floor space. The general practice is to raise meat birds on a floor with litter, 10 to 12 cm thick. The type of litter depends upon availability, suitability and economics. Good litter should be clean, dust-free, absorbent and not too fine (because chicks tend to eat small, particle-size litter). Some commonly used litter materials include wood shavings, sawdust, chopped hay, rice hulls, peanut hulls, shredded paper, shredded coconut husks, sand and chopped corncobs. For Guam and Micronesia, chopped dried *Leucaena sp* (tangantangan) branches make good litter for poultry.

Regardless of the type of litter used, avoid moldy and wet materials. Between flocks, litter can be removed completely or a layer of fresh litter can be added on top of the old litter.

<u>The following conditions are necessary whenever old litter is to be reused:</u>

1. Reuse litter with no case history of disease outbreak
2. Allow at least a three-week period between batches
3. Remove the litter in all wet spots around waterers
4. Remove all caked litter

5. Allow for proper ventilation to get rid of ammonia
6. The new litter should be at least 5 cm thick

Lighting

Broilers can be reared with 24 hours of lighting without adversely affecting production. However, it is practical to allow one or two hours of darkness so that the birds will not panic in case of a power outage. Under tropical conditions, it is advantageous to provide artificial light so that feed consumption is encouraged when ambient temperature is low (at night). All 100-watt bulbs used to provide additional heat during the first 10 days should be replaced by 60-watt bulbs or bulbs of even lower wattage, evenly distributed and raised more than two meters above the floor.

Marketing Broilers

Catching and Transport

Catching and transport are essential activities for marketing broilers, but are sources of stress and lower quality if not properly carried out. Birds should be handled gently, and chicken coops with rounded corners will reduce bruises. Do not overcrowd the birds, and transportation needs to be done during cool hours (evenings or early mornings) to avoid high mortality from heat stress.

Method of Disposing of Broilers

In order to select the best market, the producer must have sufficient information at his disposal. The relative costs of marketing for the different available markets, including transportation and death loss must be determined. These are common methods of disposing of poultry:

- Home slaughter can be very important on Guam and Micronesia, especially during village fiestas, holidays and other gatherings
- Broilers may be sold live at farmer's markets

- Back-yard poultry owners may purchase broilers for fattening
- Local butchers may purchase live broilers
- Supermarkets may purchase dressed broilers

Basis for Broiler Sales

At the market place, broilers can be sold on the following basis:

- By the head
- By live weight
- By carcass weight or yield
- By cut-out value

    Any dressed poultry designed to be sold to the public requires USDA-Approved slaughter facilities and carcass inspection. A dressed bird represents 68 to 72% of live weight.

Marketing Costs of Broilers

Charges or costs involved in marketing poultry depend on the type of services and facilities used, and how the final product is sold (on a live or dressed basis). Here are some costs involved in marketing broilers:

- Advertising
- Transportation
- Part-time labor (catching crew)
- Death loss (during transport and hauling)
- Refrigeration
- Slaughtering and packaging
- Feeding and watering (live market)

Broiler Nutrition

<u>Feeding Practices</u>

Large-scale broiler operation feeding practices are fully automated. However, small-scale farms may use manual hanging feeders with satisfactory results. A 50 lb (25 kg) feeder will provide for 70 broiler chickens. Feed should be available at all times since restricted feeding may slow the growth rate of broilers. Night feeding should be encouraged with artificial light to improve growth rate as lower environmental temperatures increase feed intake. Broiler feed may be mashed, crumbled or pelletized. Generally feed in pellet form tends to increase feed intake, reduces feed wastage and improves body weight and feed conversion. However, mashing feed with the proper amount of fat to improve palatability reduces dust and may decrease feed cost, compared to pellets. Diets can be changed from starter blend (22 to 23% protein) to a grower blend with 19 to 21% protein level at any time between 20 to 25 days of age, without adversely affecting market weight at 49 days of age (Diambra and McCartney, 1985). Ingredients high in dietary fiber will cause a reduction in growth rate of broilers. Cellulase enzyme and amino acids supplementation will improve broiler performance when leucaena leaf meal is used in the diet (Abawi and Diambra, 1994).

Organic farmers may include free range in their operation and supplement with food scrap, grocery food waste or any grain, free from molds. Modern broiler breeds will respond well to adequate feed but will display slower growth with inadequate feed and more space to go around. Water and feed must be available at all times to improve growth. Models of sustainable farming may include a rotation of the free range area with vegetable crops, or free range under fruit crops such as papaya or any other tropical fruit crop. Good fencing will protect your chickens from predators such as dogs.

The following Table describes the nutrient requirements for stages of growth.

Nutrient Requirements

Table 1a. Nutrient Requirements of Broilers

Nutrients & Energy	Units	Starter 0 to3 weeks	Grower-finisher 3 wks. to market (6-7 wks.)
Metabolizable energy	Kcal/kg	2950-3100	2950-3200
Protein	%	22-23	19-21
Calcium	%	0.90-1.10	0.90-1.10
Phosphorus (avail.)	%	0.44-0.46	0.40-0.42
Lysine	%	1.20-1.22	1.00-1.10
Methionine+cystine	%	0.92-0.94	0.70-0.72
Crude fiber	%	Max 4	Max 5

Adapted from National Research Council (1994)

Practical Feed Formula for the Tropics:

Feed for broilers is computer-formulated and varies according to the availability of ingredients, and the nutrient requirements for a given age of broilers. There is no universal feed formula for broilers, in regards to ingredients. Nutrient contributions of the ingredients are more important than the ingredients per se. Fulfilling the nutrient requirements of broilers is the cornerstone of a good diet. However, no matter what ingredients are used, they must be of good quality (fresh, dry, no mold or insect infestation, homogenous, coarsely grounded). The major contribution of each ingredient provides a guideline for ingredient substitution. The following is a list of potential feed ingredients for the Tropics.

Ingredient Major	Contribution
Beans	energy, protein, fiber
Brewer by-products	energy
Cassava, taro, yam, breadfruit	energy (starch)
Coconut, coprah	protein, fat, fiber
Coral, oyster or snail shell	calcium
Cottonseed meal	protein
Edible leaves (taro, yam)	energy, protein, fiber
Fats and oils (palm, coconut)	energy
Fishery-by-product	protein, calcium, phosphorus
Fruits (banana, papaya, jackfruit)	energy
Grain (corn, sorghum)	energy (starch)
Mill by-products (wheat, rice)	energy, fiber
Peanut meal	protein
Pumpkin seed	protein, fat
Restaurant fat and grease	energy (fat)
Restaurant food waste	energy, protein, calcium
Rice paddy	energy, fiber
Slaughter house by-product	protein, calcium, fat, phosphorus
Snail meal	protein
Soybean meal	protein
Tangantangan leaves	protein, fiber
Termites, crickets	protein

Table 1b. Practical Feed Formula for Broilers with High Level of Tangantangan

Ingredients	Starter Percent	Grower-Finisher Percent
Yellow corn	50.00	52.00
Soybean meal	28.00	20.00
Tangantangan	10.00	16.00
Fish meal	4.11	5.62
Limestone (coral)	1.30	0.82
Coconut oil	4.09	3.55
Dicalcium phosphate	1.31	0.87
Salt (NaCl)	0.50	0.37
Vitamin & mineral mix	0.50	0.50
Lysine	0.05	0.05
Methionine	0.15	0.15

Approximate Calculated Analysis

Metabolizable energy kcal/kg	2990	2990
Protein %	23.0	21.0
Calcium %	1.10	1.09
Phosphorus (avail.) %	0.45	0.50
Crude fiber %	4.0	5.0
Lysine %	1.30	1.16
Methionine+cystine	0.93	0.90

Environment and Health

<u>Ventilation</u>

Ventilation of poultry houses is necessary for the following reasons:

1. Renews/Turns over oxygen
2. Removes carbon dioxide, ammonia and other noxious gases
3. Helps to control moisture build-up from manure and water
4. Helps lower house temperature

Proper ventilation can be achieved by an appropriate orientation and design of the building or by use of forced draft ventilation.

<u>Ammonia</u>

A high ammonia concentration has the following adverse effects:

1. Increases susceptibility to respiratory diseases
2. Causes air sac infections (**airsaculitis**)
3. Induces ammonia blindness (**keratoconjunctivitis**)
4. Increases the incidence of breast blisters

Ammonia in the poultry house can be reduced by keeping the litter dry, or by adding a fresh layer of litter. Good ventilation will reduce the risk of ammonia build-up. Optimum floor density will prevent excess ammonia inside the building.

Vaccination

Consult your local veterinarian or extension agent prior to using vaccines and other medication to obtain information on the types of vaccine needed in your area.

Generally, vaccine is administered through the drinking water. Be aware that most public water contains chlorine, which could inactivate the vaccine. If a source of de-chlorinated water (rain water for example) is not available, it is recommended to add 4 to 5 teaspoons of powdered skim milk per gallon to activate the vaccine. Keep all vaccine refrigerated (not frozen) until use. Be sure to use the recommended dosage and always check for the expiration date. All vaccine containers should be discarded properly.

Before administering the vaccine via water, a rule of thumb is to remove water for an hour or two to induce rapid consumption of the vaccine. Water cups or fountains and buckets used for the water-vaccine mixture should be clean. After the vaccine is consumed, get the birds back on the normal water regimen.

A farmer should always keep a record of the date, type and dose of the vaccine administered. Whenever spray is used for vaccination, avoid drafty condition and apply the spray evenly. Due to a broiler's short life span (6 to 7 weeks), the bird may not need as much vaccine as pullets or layers. It may be useful to vaccinate against Newcastle, bronchitis and infectious bursal disease. Flocks under 500 birds may not need to be vaccinated, but then good hygiene becomes critical.

Major Broiler Health Concern

Coccidiosis

Primarily a disease of the digestive tract, coccidiosis is caused by a protozoa called coccidian, characterized by bloody diarrhea. The intestinal tract is affected and the area around the vent is stained with blood. There are several species of coccidian, but 3 species severely affect chickens. Table 2a shows regions of the digestive tract affected and types of lesions observed.

Table 2a. Three prevalent species of coccidian affecting chickens

Species of coccidian	Part of digestive tract affected	Lesions observed
E. acervulina[1]	Upper small intestine	Whitish patches, thickened intestine wall
E. tenella	Ceca	Hemorrhaging
E. necatrix	Small intestine	Ballooning, hemorrhaging

[1]E stands for *Eimeria*
Source: *Salsbury Manual of Poultry Diseases (1979)*

A wide range of anticoccidial drugs (Monensin, Salinomycin, Lasalocid) is available for treatment via drinking water or feed. To avoid residue in the carcass, a withdrawal period of 3 to 7 days (according to the specific drug) is to be strictly observed before slaughter. Immunox provides life-time immunity against main pathogenic coccidian.

Internal and External Parasites

There are numerous species of roundworms, tapeworms and capillary worms known to occur in poultry. They are not a serious concern in broiler chickens due to the short lives of the hosts not allowing time for most parasites to complete their life cycle. Flubernol and Aviverm are effective against roundworms and tapeworms.

Most ectoparasites (lice, mites, ticks) live on skin or feathers. Chemicals like Coumaphos, Malathion and Sevin, when carefully used, are effective in infested buildings. Consult with your local veterinarian or extension agent for more details.

Coli Infection

"Coli" refers to *Escherichia coli*, bacteria commonly found in the intestinal tract of birds, animals and man, and in the environment. Coli infection in chickens may take the form of a hemorrhaging infection, an air-sac infection or a localized infection in any tissue of the body, producing an inflammation. Diseases or stress conditions may predispose a chicken to *E. coli* infection. Certain strains of *E. coli* can produce high mortality and condemnation of carcasses. Correctly diagnosing an *E. coli* infection can be very difficult since lesions in many organs may be caused by organisms other than *E. coli*. This infection is a constant threat since chickens may be contaminated through droppings, feed, water, litter, dust, air, equipment, people, wild birds, rodents and insects. Moreover, a chick embryo may be contaminated through the egg shell.

The best prevention against *E. coli* infection is a strict sanitation program. *E. coli* infections are treated with antibiotics, which may be administered via water or feed during stress (heat stress or vaccination). However, as birds may develop resistance to repeated antibiotics, it is always best to consult with a local veterinarian or extension agent for treatment.

Salmonella Infections

Salmonella infections have become a great concern in the field of public health because about 35 types are known to have been transmitted to humans; the human form induces an upset stomach. The mortality rate in chicks varies widely (2 to 50%). The rate of spread of the disease and mortality rate are often influenced by environmental factors such as chilling, overcrowding, unsanitary conditions, faulty ventilation and other disease conditions.

Chicks dying soon after hatching often exhibit no sign before death, but older chicks may appear exhausted and sleepy. There is usually loss of appetite, evidence of diarrhea and **pasting** around the vent. Generally the disease appears at two different age periods: first, when birds are four to five days old; then, when the birds are 10 to 12 days old.

Treatment with furazolidone at a concentration of 0.04% in the diet of chicks for a period of 10 days can be highly effective. This drug can also be used as a preventative measure in starter ration at 0.01%.

Heat Stress

The tropical environment makes broilers very susceptible to heat stress, especially in periods of high ambient temperature and high humidity. Having no sweat glands, a bird can lower its body temperature only by rapid respiration with the mouth wide open. High mortality may occur due to heat stress, especially when birds reach market size, which may result in high economic losses.

Heat stress can be prevented by installing mechanical ventilation, maintaining proper floor density and providing sufficient shade around the poultry house. Dry litter will help keep humidity down in the poultry house.

Birds on free range need ample shade to avoid direct exposure to sunlight. Fresh water available at all times will help birds stay cool.

Infectious Bursal Disease (IBD or Gumboro)

IBD is an important viral disease affecting poultry around the world. Young birds between three and six weeks of age are most often affected. Following infection there is a short incubation period (two days) before the bursa becomes congested and swollen. In the acute form, a whitish and watery diarrhea is observed. There is a lack of appetite and falling posture. Deaths occur within two days of the first signs of the disease and reach a peak in two or three more days then rapidly decline. Carcasses of birds dying from this infection are dehydrated.

Good sanitation will limit the spread of the disease. Vaccination is highly recommended because there is no treatment for IBD.

Newcastle Disease (NCD)

In an outbreak of NCD, mortality may be as high as 50% of the flock. The viral disease is known in almost every area of the world. Signs of the disease are gasping, coughing, loss of appetite and nervous symptoms including partial or complete paralysis of the legs or wings. A typical posture includes letting the head droop down between the legs, or letting the head fall straight back between the shoulders, rotating of head and neck, walking backwards, circling. All chickens should be vaccinated as there is no treatment for NCD.

Infectious Bronchitis (IB)

An acute viral disease affecting chickens of all ages, IB is probably the most widespread respiratory disease. Respiratory discomfort is manifested both in young chickens and older birds by gasping and coughing. Nasal discharge is common in young birds, along with wet eyes. A drop in feed consumption follows immediately.

Both IB and NCD affect the respiratory tract. To differentiate IB from NCD, observe a lack of nervous symptoms with IB. There is no treatment for IB and vaccination will protect the flock.

Nutritional Deficiency Diseases

Nutritional deficiency diseases affect a small proportion of broiler flock (less than 1%) when the farmer feeds a balanced ration. However, signs of nutritional deficiency may be observed when some external factors prevent normal feed consumption or normal nutrient absorption. The first signs of any deficiency are usually retarded growth and poor feathering. Table 2b shows common deficiency symptoms of mineral and vitamins.

Table 2b. Common Nutritional Deficiency Symptoms (1)

Symptoms	Deficiency Compounds
Rickets (skeletal deformities)	Calcium, phosphorus, vitamin D3
Perosis (crippled leg or slipped tendon)	Manganese, choline, biotin
Poor feathering	Zinc, folic acid
Crazy chick disease(uncoordinated movement)	Vitamin E
Curled-toe paralysis	Vitamin B2
Intramuscular bleeding	Vitamin K

Source: Gordon & Jordan (1982)

How to Recognize Symptoms of a Disease Condition

Respiratory—coughing, sneezing, head-shaking, swollen eyes, nasal discharge

Nervous—trembling, falling, paralysis, circling, blindness

Locomotion—crooked toes, enlarged hocks, slipped tendon, paralysis, bone conformation, posture

External—poor feathering, skin or eye lesions, parasites

Droppings—watery consistency, bloody, white, green

Conditions which induce stress factors—moving, overcrowding, vaccination, drug treatment, high temperature and humidity, night chilling, lack of water or feed, drastic change of feed, excessive noise

Practical Recommendations for Sanitation

To use good sanitation practices, farm isolation is the best insurance against disease affecting the flock. Routine cleanout practices, coupled with good ventilation, can help prevent susceptible birds from being infected. Between flocks, the following are recommended:

1. Wash the entire poultry house to eliminate all dust and debris
2. **Disinfect** all surfaces with approved disinfecting agents
3. Spread new litter to a depth of 10 to 12 cm
4. Make sure litter is clean, dust-free, dry, absorbent, economical and easily available. Too fine a material will encourage chicks to eat litter instead of feed.
5. Clean and disinfect all feeding and watering equipment between flocks and routinely as needed.

Waste Management

Manure

Most commercial operations raise broilers on dry litter floor where manure and bedding are mixed. The quality of the mixture will vary with the bedding mixture ratio, age of litter, type of bird, type of feed and type of bedding. Broiler litter is a desirable soil amendment and produces beneficial effects on soil properties (Motavalli and Diambra, 1997). However, N leaching is a concern for groundwater contamination. Recycling dehydrated broiler litter by re-feeding has been demonstrated to be economically and nutritionally beneficial for ruminant animals (goat, sheep, cow, water buffalo). The Inarajan Agricultural Experiment Station of the University of Guam has established proper conditions for ensiling poultry manure as a feed supplement for ruminants. The broiler-grower should take steps to convert poultry waste into a valuable resource such as soil amendment or ruminant feed supplement. Farmers growing crops will usually utilize poultry manure mixed with litter for organic farming. Therefore, bagging manure and litter provides additional revenue to poultry growers.

Broilers raised on free range have less need to recycle manure. When combined with a crop rotation system or raised under shrubs, the manure is directly incorporated into the soil.

Disposal of Dead Birds

Dead bird disposal may become a serious concern for a farm size above 5,000 broilers, giving an average flock mortality of 5%. Dead birds may be buried, incinerated or composted. Composting dead birds has the advantage of returning the compost material to the soil. Composting dead birds is achieved by making several layers of dead bird, poultry litter and straw above a concrete floor covered with the straw, into a square pit. Different designs of composting bins are available. More information can be obtained from your extension agent. Do not leave dead birds exposed.

Miscellaneous Tables

The following tables provide the grower with some benchmarks to keep the broiler operation on tract.

Table 3: Straight Run Broiler Performance (day old body weight: 38g)

Age	Body Weight	Feed Consumption	Cumulative Feed Consumption	Feed Conversion
Days	Average, g	Daily Average, g	Weekly Cumulative	
7	115	15.4	108	1.40
14	274	34.2	347	1.50
21	550	67.8	822	1.71
28	870	82.8	1402	1.81
35	1220	96.2	2075	1.92
42	1630	114.6	2877	1.95
49	2040	120.8	3722	2.06

Source: Inarajan Agricultural Experiment Station, the University of Guam (Broiler chicks from Asagi Hatchery, Honolulu, Hawaii)

Table 4: Daily Water Consumption

Age of Broiler (weeks)	Liters per 1000 Birds/Day
1	2-3
2	4-5
3	6-7
4	8-9
5	10-11
6	12-13
7	14-15
8	16-17

Source: Gordon & Jordan (1982)

As a "rule of thumb," water consumption in liters per 100 birds is estimated by multiplying the age in weeks by 2.1.

Table 5: Cost of Production (Template)[1]

Items	% of cost
Feed	
Stock	
Medication and Vaccination	
Fuel	
Other (litter, electricity, water, waste handling)	
Fixed cost	
Total variable cost	
Grower payment	
Total	

[1] To be completed by poultry grower

Table 6: Components of a Broiler (1.86 kg live weight)

Components	Grams	% Live Weight
Blood	77.2	4.14
Feathers	151.0	8.10
Head	44.1	2.37
Feet	72.6	3.90
Total offal	344.9	18.51
Intestinal tract	149.0	8.00
Lung, spleen, pancreas	16.8	0.90
Kidneys	132.2	0.66
Total viscera	178	9.56
Heart	9.0	0.49
Liver	39.0	2.10
Gizzard	55.4	3.00
Neck	77.7	4.20
Eviscerated carcass	1158	62.20
Giblets	182	9.80
Total saleable	1330	72.00

Source: Cobb broiler manual (1983)

References

Abawi, F. G., & Diambra, O. H. (1994). Effect of enzyme and amino acid supplementation of leucaena leaf meal diet on broilers. *Poultry Sci.*, *73*, 95.

Cobb broiler manual (1983). Concord (Massachusetts): Cobb Inc.

Diambra, O. H., & McCartney, M. G. (1985). Performances of male broilers changed from starter to finisher diets at different ages. Poultry Sci., 64, 1829-33.

Gordon, R. F., & Jordan, F. T. W. (1982). *Poultry diseases.* London: Bailliere Tindal.

Inarajan Agricultural Experiment Station. (1995). *Experiment data on poultry production. Compounded poultry data. 1993-1995.*

Motavalli, P. P. and Diambra, O. H. (1997). Management of Nitrogen Immobilization from Waste Office Paper Applications to Tropical Pacific Island Soils. Compost Science & Utilization, Vol. 5, No. 3.

National Research Council. (1994). *Nutrient requirements of poultry.* Washington D.C.: National Academy Press.

Salsbury Manual of Poultry Diseases. (1979). Charter City (Iowa): Salsbury Laboratories.

Management Records (Template)

farm name _____   date started _____

number started _____   origin of chicks _____

week	1	2	3	4	5	6	7	cumul.
no. of feed bags (bag wt = --)								
no. of dead birds								
av. body wt. of 20 birds								

<u>Medication</u>

coccidiostat name _____   date given _____   withdrawal _____

antibiotic name _____   date given _____   withdrawal _____

other _____   date given _____   withdrawal _____

<u>Vaccination</u>

type of vaccine _____ date given _____ via _____

_____   _____   _____

_____   _____   _____

number of birds sold _____date sold _____

number of birds culled (retarded growth, crooked legs . . .) _____

estimated **feed conversion** = <u>total amount of feed</u>

                                    total amount of live weight

other comments _____

_____

_____

Conversion Table

1 US ounce	= 28.35 grams
1 pound	= 454 grams
	= 0.454 kilogram
10.76 sq ft.	= 1 sq meter
100 sq ft.	= 9.3 sq meters
1 sq in.	= 6.452 sq cms
1 sq yd.	= 0.836 sq meter
1 gallon	= 3.784 liters
(°F - 32) x 5/9	= °C
(°C x 9/5) + 32	= F

Glossary

all-in/all-out: a management practice whereby birds are raised at the same age and sold all at once.

airsaculitis: an infection caused by *Escherichia coli* affecting the air sacs of birds.

broiler: meat-type chicken, also called fryer.

brooder: an area where chicks are provided an extra heat source.

disinfect: to kill or otherwise render ineffective harmful microorganisms and parasites.

feed conversion: a measure of feed-efficiency expressed as the ratio between feed consumed and unit of production. For broilers, feed conversion is expressed as kg of feed per kg of body weight or kg of feed per kg of weight gain for a given period.

floor density: number of birds per surface area.

keratoconjunctivitis: also called ammonia blindness, and caused by excessive ammonia creating an eye infection.

pasting: digestive upset that results in sticky droppings adhering to the vent area eventually plugging the intestine and causing death.

straight run: a flock of broilers of mixed sexes.

Brooding in cage with manual water cup. *(source: Manny Duguies)*

Brooding in cage with automatic water cup

Brooding inside a container with straw litter

Broilers ready for market or slaughter house *(source: Hauhouot Diambra-Odi)*

Type of broiler housing for the tropics *(source: Hauhouot Diambra-Odi)*

CHICKEN RECIPES

Chicken Keleguen (Guam's favorite)

Ingredients:

- 2 roasted (or barbecued) broiler chicken breast or one whole boneless chicken
- 1/3 cup of lemon powder or fresh lemon juice
- 1/2 cup of green onions
- 1/2 tablespoon of crushed garlic
- 1 tablespoon of crushed onions
- 2 tablespoons of chopped red pepper
- 1/2 cup of grated dried coconut
- 1/4 teaspoon salt

Directions:

Cut the roasted boneless chicken breast after removing the skin, in very small pieces (or coarsely chopped in a blender), and mix with lemon powder or fresh lemon juice in a bowl. Add crushed garlic, crushed onions, chopped red pepper and grated coconut meat. Add salt and chopped green onions. Mix well and refrigerate for 30 minutes. Serve cold as an appetizer.

Serving size: for four people

Source: Julius Naranjo and Kateri Cruz (2013)

Chicken Motsiyas (Guam Specialty)

Ingredients:

- 2 broiler chicken breasts or one whole boneless chicken
- 1/2 cup of long beans
- 1/2 cup of wing beans
- 1/2 cup of young pumpkin tips
- One whole onions
- 1/2 tablespoon of garlic
- 1/4 cup of lemon powder or fresh lemon juice
- Pumpkin leaves (for wrapping)
- One cup of coconut milk
- One tablespoon of chopped red pepper
- 1/4 teaspoon of salt
- Vein of coconut leaves (for tying)

Directions:

Debone a whole chicken (or use two chicken breasts), coarsely chop the meat, mix with chopped vegetables (long beans, wing beans, pumpkin tips, onions, garlic); add lemon powder or fresh lemon juice and mix well inside a bowl. Take 1/3 cup of the meat and vegetable mixture with lemon and wrap into two to three pumpkin leaves tied with coconut leaf veins. Place in a cooking pot and pour coconut milk on top. Simmer for 30 to 45 minutes. Serve warm with rice, boiled cassava, boiled taro or boiled cooking bananas.

Serving size: for four people

Source: Frank M. Paulino (2013)

Chicken Yassa (Senegal's favorite)

Ingredients;

- One whole broiler chicken
- 1/2 cup of fresh lime juice
- 2 lemons (sliced)
- 3 to 4 whole onions (a mixture of white, yellow and red onions) sliced
- 1/4 cup of green onions
- 2 tablespoons of chopped garlic
- 4 to 5 dried bay leaves
- 1 teaspoon of crushed black pepper
- One tablespoon of crushed fresh ginger
- One tablespoon of fresh parsley
- One cup of sliced bell pepper (a mixture of green, yellow and red)
- 1/4 cup of vegetable oil
- One tablespoon of red pepper
- 1/4 teaspoon of salt

Directions:

Cut the chicken with bones in pieces (eight to twelve pieces), and add the lime juice and the sliced lemon in a bowl. Add the sliced onions, garlic, bell peppers, ginger and bay leaves. Let marinate in the refrigerator from twelve hours to overnight in a tight container. The next day, grill or barbecue the chicken until it turns brown. Heat vegetable oil in a cooking pan and add the marinated onions only from the mixture and cook from eight to twelve minutes first. Then add the grilled (or barbecued) chicken and the rest of the marinated mixture and let it simmer for ten minutes. Before serving, garnish with black pepper, fresh parsley and green onions. Serve warm with rice, "attieke" or couscous.

Serving size: four to six people

Source: Marcelle Diambra-Odi (2013)

Chicken Kedjenou (Cote d'Ivoire's favorite)

Ingredients:

- One whole chicken
- 4 fresh tomatoes (large)
- 2 eggplants
- One whole onion
- 1/2 tablespoon of crushed garlic
- 5 dried bay leaves
- 2 "Maggi" cubes
- One yellow bell pepper
- One cup of spinach or sweet potato leaves
- One cup of squash
- 2 tablespoons of tomato paste
- 1/4 cup of cooking wine (optional)

Directions:

Cut chicken in bite-sizes, chop all vegetables, and add "Maggi" cubes, bay leaves, tomato paste and cooking wine into a cooking clay pot tightly covered with banana leaves or aluminum foil. No water is added. Simmer for 30 to 40 minutes. Serve with "attieke", boiled cassava, boiled cooking banana, boiled taro or yam, fonio couscous, wheat couscous or rice.

Serving size: four to six people

Source: Hauhouot Diambra-Odi (2013)